Preface

Welcome to the "Science Olympiad All Chapters Combined MCQ Class 7"

This book is specially crafted to ignite the spark of curiosity and foster a love for science in young minds. Designed with the unique needs of first graders in mind, it provides a comprehensive introduction to fundamental scientific concepts through engaging, age-appropriate content.

Our primary goal is to make learning science an enjoyable and enriching experience. The book is filled with colorful illustrations, real-life examples, and interactive exercises that help children understand and relate to the world around them. Each chapter is carefully structured to build on prior knowledge, ensuring a steady progression in learning.

Key features of this book include:

Interactive Exercises: Hands-on activities and simple experiments designed to engage students actively and make learning fun.

Multiple Choice Questions: End-of-chapter MCQs help reinforce learning and provide ample practice for young learners.

Cross-Curricular Connections: Integrates elements of math and language arts, promoting a holistic learning experience.

Environmental Awareness: Introduction to basic environmental science and sustainability concepts.

Through this book, we aim to encourage critical thinking, problem-solving skills, and a deeper understanding of scientific principles. The inclusion of real-life scientists and their contributions will inspire young learners and highlight the importance of science in our daily lives.

Whether used in the classroom or at home, this book is an invaluable resource for fostering a lifelong love of science. We hope that it will be a stepping stone for students, helping them explore and appreciate the wonders of the scientific world.

Thank you for choosing our Science Olympiad Book. Let's embark on this fascinating journey of discovery and learning together!

Happy Learning!

Table of Contents

Questions: Part 1

1. Which part of the plant absorbs water and minerals from the soil?

   a) Stem

   b) Leaf

   c) Root

   d) Flower

2. What is the process by which green plants make their food?

   a) Respiration

   b) Transpiration

   c) Photosynthesis

   d) Germination

3. Which nutrient is primarily responsible for building muscle in animals?

   a) Fats

   b) Carbohydrates

   c) Proteins

   d) Vitamins

4. What is the process called where animals break down food to release energy?

   a) Digestion

   b) Respiration

   c) Circulation

   d) Excretion

5. Which natural fiber is obtained from sheep?

   a) Cotton

   b) Silk

   c) Wool

d) Jute

6. Which of the following is an example of a synthetic fiber?

a) Cotton

b) Wool

c) Polyester

d) Silk

7. What is the measure of the average kinetic energy of the particles in a substance?

a) Temperature

b) Heat

c) Cold

d) Energy

8. What is the method of heat transfer in liquids and gases?

a) Conduction

b) Convection

c) Radiation

d) Absorption

9. What is the pH value of a neutral solution?

a) 0

b) 7

c) 14

d) 5

10. Which of these is an example of a strong acid?

a) Vinegar

b) Lemon juice

c) Hydrochloric acid

d) Milk

11. Which of the following is a physical change?

a) Rusting of iron

b) Burning of wood

c) Melting of ice

d) Cooking of food

12. What kind of change involves the formation of a new substance?

a) Physical change

b) Chemical change

c) No change

d) Reversible change

13. What is the term for the long-term average of weather patterns in an area?

a) Weather

b) Climate

c) Storm

d) Wind

14. What is the primary cause of global warming?

a) Greenhouse gases

b) Solar radiation

c) Earth's tilt

d) Ocean currents

15. Which of these animals is well adapted to a cold climate?

a) Polar bear

b) Elephant

c) Camel

d) Lion

16. What device is used to measure wind speed?

a) Thermometer

b) Barometer

c) Anemometer

d) Hygrometer

17. What causes cyclones?

a) High-pressure systems

b) Low-pressure systems

c) Equal pressure systems

d) Cold fronts

18. Which of these is not a type of soil?

a) Sandy soil

b) Clay soil

c) Rocky soil

d) Loamy soil

19. What is the main component of soil?

a) Minerals

b) Water

c) Organic matter

d) Air

20. Which process in plants involves the intake of oxygen and release of carbon dioxide?

a) Photosynthesis

b) Respiration

c) Transpiration

d) Germination

21. Which part of the human respiratory system is responsible for gas exchange?

a) Trachea

b) Bronchi

c) Alveoli

d) Diaphragm

22. What transports water and nutrients from the roots to other parts of the plant?

a) Phloem

b) Xylem

c) Cortex

d) Pith

23. Which blood component carries oxygen throughout the body?

a) White blood cells

b) Red blood cells

c) Platelets

d) Plasma

24. What is the process of producing offspring called in plants?

a) Pollination

b) Germination

c) Reproduction

d) Fertilization

25. What is the male reproductive part of a flower called?

a) Stigma

b) Ovary

c) Stamen

d) Petal

26. Which part of a plant develops into a fruit?

a) Leaf

b) Stem

c) Ovary

d) Root

27. What is the unit of speed?

a) Meter

b) Second

c) Kilometer

d) Meter per second

28. What is the term for the change in position of an object over time?

a) Speed

b) Motion

c) Distance

d) Displacement

29. What is the SI unit of time?

a) Minute

b) Hour

c) Second

d) Day

30. Which of these is a source of direct current (DC)?

a) Battery

b) Power plant

c) Generator

d) Solar panel

31. What is the effect of a current passing through a wire wrapped around a magnetic material?

a) It becomes demagnetized

b) It becomes a magnet

c) It melts

d) It evaporates

32. What happens when two like charges are brought close together?

a) They attract

b) They repel

c) They neutralize

d) They fuse

33. What is the main source of light on Earth?

a) The Moon

b) The Stars

c) The Sun

d) Fire

34. What type of image is formed by a plane mirror?

a) Real and inverted

b) Real and upright

c) Virtual and inverted

d) Virtual and upright

35. Which of these is an example of a renewable resource?

a) Coal

b) Oil

c) Wind

d) Natural gas

36. What process describes the movement of water from the ground to the atmosphere?

a) Condensation

b) Evaporation

c) Precipitation

d) Infiltration

37. What percentage of the Earth's water is fresh water?

a) 50%

b) 97%

c) 2.5%

d) 70%

38. What is the primary function of forests in the environment?

a) To produce oxygen

b) To absorb sunlight

c) To reflect heat

d) To create storms

39. Which layer of the soil is richest in humus?

a) Topsoil

b) Subsoil

c) Parent rock

d) Bedrock

40. What term is used for water that is not fit for human consumption due to pollution?

a) Potable water

b) Wastewater

c) Groundwater

d) Mineral water

41. Which of the following nutrients is essential for plant growth?

a) Nitrogen

b) Oxygen

c) Carbon dioxide

d) Hydrogen

42. Which vitamin can be synthesized by the human body in the presence of sunlight?

a) Vitamin A

b) Vitamin B

c) Vitamin C

d) Vitamin D

43. What process is responsible for the rise of water in plants through xylem vessels?

a) Capillary action

b) Diffusion

c) Osmosis

d) Active transport

44. What is the scientific study of soil called?

a) Ecology

b) Botany

c) Pedology

d) Zoology

45. What is the role of stomata in leaves?

a) To absorb water

b) To exchange gases

c) To store nutrients

d) To produce chlorophyll

46. Which of the following is a physical change?

a) Burning wood

b) Freezing water

c) Rusting iron

d) Baking a cake

47. What is the SI unit of electric current?

a) Volt

b) Watt

c) Ohm

d) Ampere

48. What phenomenon explains why a pencil appears bent when placed in water?

a) Reflection

b) Refraction

c) Diffraction

d) Dispersion

49. What is the force that pulls objects toward the Earth?

a) Magnetic force

b) Gravitational force

c) Electromagnetic force

d) Nuclear force

50. Which of the following is a chemical change?

a) Boiling water

b) Melting ice

c) Burning paper

d) Dissolving sugar in water

Questions: Part 2

51. What type of energy is stored in a stretched rubber band?

a) Kinetic energy

b) Potential energy

c) Thermal energy

d) Chemical energy

52. What is the function of the nucleus in a cell?

a) Produces energy

b) Controls cell activities

c) Provides structure

d) Protects the cell

53. Which of the following is an example of a nonrenewable resource?

a) Solar energy

b) Wind energy

c) Natural gas

d) Geothermal energy

54. What is the process by which plants lose water through their leaves?

a) Photosynthesis

b) Transpiration

c) Respiration

d) Germination

55. Which part of the brain is responsible for memory?

a) Cerebrum

b) Cerebellum

c) Medulla

d) Pons

56. Which process leads to the formation of clouds?

a) Precipitation

b) Evaporation

c) Condensation

d) Sublimation

57. What is the chemical formula for water?

a) H_2O

b) CO_2

c) O_2

d) H_2SO_4

58. Which gas is produced by burning fossil fuels?

a) Oxygen

b) Nitrogen

c) Carbon dioxide

d) Hydrogen

59. What type of soil is best for growing crops?

a) Sandy soil

b) Clay soil

c) Loamy soil

d) Rocky soil

60. What is the function of chlorophyll in plants?

a) Absorbs water

b) Absorbs sunlight

c) Stores nutrients

d) Protects the plant

61. Which device converts mechanical energy into electrical energy?

a) Generator

b) Battery

c) Transformer

d) Capacitor

62. Which part of the human skeleton protects the brain?

a) Rib cage

b) Pelvis

c) Skull

d) Spine

63. What is the main cause of acid rain?

a) Greenhouse gases

b) Ozone depletion

c) Emission of sulfur dioxide and nitrogen oxides

d) Increased carbon dioxide levels

64. Which of the following is a renewable source of energy?

a) Coal

b) Oil

c) Natural gas

d) Solar energy

65. Which gas do plants absorb from the atmosphere for photosynthesis?

a) Oxygen

b) Nitrogen

c) Carbon dioxide

d) Methane

66. What is the primary function of red blood cells?

a) Fight infections

b) Carry oxygen

c) Clot blood

d) Digest food

67. What is the process by which rocks are broken down into smaller pieces?

a) Erosion

b) Weathering

c) Sedimentation

d) Deposition

68. Which part of the human body is responsible for pumping blood?

a) Lungs

b) Brain

c) Heart

d) Liver

69. What is the main function of roots in a plant?

a) Produce flowers

b) Absorb water and nutrients

c) Store food

d) Perform photosynthesis

70. Which of the following materials is a good conductor of electricity?

a) Rubber

b) Plastic

c) Copper

d) Wood

71. What type of motion does a pendulum exhibit?

a) Circular motion

b) Rectilinear motion

c) Oscillatory motion

d) Random motion

72. Which of these is not a characteristic of living organisms?

a) Growth

b) Reproduction

c) Inertia

d) Respiration

73. What is the primary source of energy for the Earth's climate system?

a) Wind

b) Ocean currents

c) The Sun

d) Volcanoes

74. What is the process by which plants produce their own food?

a) Respiration

b) Photosynthesis

c) Transpiration

d) Digestion

75. Which part of the flower contains the ovules?

a) Stigma

b) Anther

c) Ovary

d) Petal

76. Which of the following is a product of photosynthesis?

a) Carbon dioxide

b) Water

c) Oxygen

d) Nitrogen

77. What is the chemical symbol for sodium?

a) S

b) So

c) Na

d) N

78. What type of change occurs when ice melts into water?

a) Chemical change

b) Physical change

c) Reversible change

d) Irreversible change

79. Which of the following is an omnivore?

a) Cow

b) Lion

c) Bear

d) Deer

80. What instrument is used to measure temperature?

a) Barometer

b) Thermometer

c) Anemometer

d) Hygrometer

81. What is the SI unit of force?

a) Joule

b) Newton

c) Watt

d) Ampere

82. What is the role of platelets in the blood?

a) Carry oxygen

b) Fight infection

c) Clot blood

d) Remove waste

83. What is the phenomenon of bending of light when it passes from one medium to another?

a) Reflection

b) Refraction

c) Dispersion

d) Diffraction

84. Which of the following processes is involved in the water cycle?

a) Photosynthesis

b) Evaporation

c) Erosion

d) Sedimentation

85. What is the name of the process by which plants lose water through small openings on their leaves?

a) Respiration

b) Transpiration

c) Photosynthesis

d) Germination

86. Which of these is a fossil fuel?

a) Solar energy

b) Wind energy

c) Natural gas

d) Geothermal energy

87. What is the process of changing liquid water into water vapor?

a) Condensation

b) Evaporation

c) Precipitation

d) Infiltration

88. What is the function of the large intestine in the digestive system?

a) Absorb nutrients

b) Store bile

c) Absorb water and form feces

d) Digest proteins

89. What is the smallest unit of life?

a) Atom

b) Molecule

c) Cell

d) Tissue

90. What is the primary gas found in Earth's atmosphere?

a) Oxygen

b) Carbon dioxide

c) Nitrogen

d) Hydrogen

91. What is the term for the variety of life in a particular habitat or ecosystem?

a) Biodiversity

b) Ecology

c) Evolution

d) Habitat

92. Which part of the cell contains the genetic material?

a) Cytoplasm

b) Nucleus

c) Cell membrane

d) Mitochondria

93. What is the process by which plants use sunlight to make food?

a) Transpiration

b) Photosynthesis

c) Respiration

d) Digestion

94. Which of the following is an example of a conductor of heat?

a) Wood

b) Plastic

c) Copper

d) Rubber

95. What is the basic unit of matter?

a) Cell

b) Atom

c) Molecule

d) Electron

96. Which type of soil has the highest water retention capacity?

a) Sandy soil

b) Clay soil

c) Loamy soil

d) Gravel

97. What is the role of the chloroplast in plant cells?

a) Store nutrients

b) Produce energy

c) Perform photosynthesis

d) Support the cell structure

98. What is the primary reason for soil erosion?

a) Heavy rainfall

b) Strong winds

c) Deforestation

d) Volcanic eruptions

99. What is the process by which seeds sprout and begin to grow into a new plant?

a) Photosynthesis

b) Germination

c) Transpiration

d) Pollination

100. Which of the following is a simple machine that helps lift objects?

a) Lever

b) Pulley

c) Wheel and axle

d) Inclined plane

Questions – Part 3

101. Which of the following is a physical change?

a) Burning of paper

b) Rusting of iron

c) Melting of ice

d) Cooking of food

102. Which gas is produced during respiration in humans?

a) Oxygen

b) Nitrogen

c) Carbon dioxide

d) Hydrogen

103. Which part of the plant conducts food from leaves to other parts?

a) Xylem

b) Phloem

c) Cambium

d) Stomata

104. Which type of reproduction involves only one parent?

a) Sexual

b) Asexual

c) Cross-pollination

d) Fertilization

105. What is the SI unit of force?

a) Watt

b) Joule

c) Newton

d) Pascal

106. Which is the largest gland in the human body?

a) Pancreas

b) Liver

c) Kidney

d) Thyroid

107. Which gas is essential for photosynthesis?

a) Oxygen

b) Carbon dioxide

c) Nitrogen

d) Helium

108. Which part of the digestive system absorbs water?

a) Stomach

b) Small intestine

c) Large intestine

d) Liver

109. Which of these is a good conductor of electricity?

a) Rubber

b) Plastic

c) Copper

d) Wood

110. Which process is used to separate salt from sea water?

a) Sedimentation

b) Filtration

c) Distillation

d) Decantation

111. Which part of the human eye controls the amount of light entering it?

a) Cornea

b) Lens

c) Iris

d) Retina

112. Which gas is produced during the fermentation of sugar?

a) Oxygen

b) Carbon dioxide

c) Methane

d) Nitrogen

113. Which vitamin is known as ascorbic acid?

a) Vitamin A

b) Vitamin B

c) Vitamin C

d) Vitamin D

114. What is the function of platelets in blood?

a) Transport oxygen

b) Fight infection

c) Help in clotting

d) Carry nutrients

115. Which is the fastest mode of heat transfer?

a) Conduction

b) Convection

c) Radiation

d) Reflection

116. Which type of mirror is used in vehicles to see a wider area?

a) Plane mirror

b) Concave mirror

c) Convex mirror

d) None of these

117. Which is the green pigment in leaves?

a) Carotene

b) Chlorophyll

c) Xanthophyll

d) Anthocyanin

118. Which part of the plant develops into fruit?

a) Root

b) Stem

c) Ovary

d) Leaf

119. Which gas turns lime water milky?

a) Oxygen

b) Carbon dioxide

c) Hydrogen

d) Nitrogen

120. Which is the hardest natural substance?

a) Gold

b) Diamond

c) Iron

d) Steel

121. Which energy is possessed by a moving body?

a) Potential energy

b) Kinetic energy

c) Heat energy

d) Chemical energy

122. Which process converts liquid into gas without boiling?

a) Freezing

b) Condensation

c) Evaporation

d) Melting

123. Which planet is known as the "Morning Star"?

a) Mars

b) Venus

c) Mercury

d) Jupiter

124. What is the unit of electric current?

a) Volt

b) Ampere

c) Watt

d) Ohm

125. Which acid is found in lemon?

a) Tartaric acid

b) Lactic acid

c) Citric acid

d) Acetic acid

126. Which part of the human ear helps in maintaining balance?

a) Cochlea

b) Eardrum

c) Semicircular canals

d) Ear canal

127. Which is the largest planet in the solar system?

a) Earth

b) Saturn

c) Jupiter

d) Neptune

128. Which gas is known as laughing gas?

a) Nitrogen dioxide

b) Nitrous oxide

c) Nitric oxide

d) Carbon monoxide

129. Which process is used to produce curd from milk?

a) Fermentation

b) Evaporation

c) Distillation

d) Filtration

130. What type of energy is stored in food?

a) Heat energy

b) Chemical energy

c) Kinetic energy

d) Potential energy

131. Which organ in humans filters blood to remove waste?

a) Liver

b) Kidney

c) Lungs

d) Heart

132. Which type of force is exerted by a magnet?

a) Contact force

b) Non-contact force

c) Muscular force

d) Friction

133. Which of the following is NOT a renewable source of energy?

a) Solar

b) Wind

c) Coal

d) Water

134. Which part of the plant is involved in transpiration?

a) Stem

b) Roots

c) Leaves

d) Flowers

135. Which metal is liquid at room temperature?

a) Mercury

b) Aluminium

c) Copper

d) Gold

136. Which is the smallest bone in the human body?

a) Femur

b) Stapes

c) Ulna

d) Radius

137. Which process is used to kill harmful microorganisms in milk?

a) Filtration

b) Pasteurization

c) Boiling

d) Distillation

138. Which organ produces bile in humans?

a) Liver

b) Pancreas

c) Gall bladder

d) Stomach

139. Which planet has the maximum number of moons?

a) Jupiter

b) Saturn

c) Neptune

d) Uranus

140. Which is the fastest bird in the world?

a) Eagle

b) Peregrine falcon

c) Owl

d) Hawk

141. Which device is used to measure atmospheric pressure?

a) Thermometer

b) Barometer

c) Hygrometer

d) Anemometer

142. Which type of lens is used in magnifying glass?

a) Concave lens

b) Convex lens

c) Cylindrical lens

d) Plano lens

143. Which gas do we inhale while breathing?

a) Oxygen

b) Carbon dioxide

c) Nitrogen

d) Helium

144. Which process involves the conversion of solid directly into gas?

a) Sublimation

b) Evaporation

c) Condensation

d) Freezing

145. Which organelle is known as the powerhouse of the cell?

a) Nucleus

b) Ribosome

c) Mitochondria

d) Chloroplast

146. Which planet is farthest from the Sun?

a) Uranus

b) Neptune

c) Saturn

d) Pluto

147. Which part of the plant stores food in potato?

a) Root

b) Stem

c) Leaf

d) Flower

148. Which gas is used in fire extinguishers?

a) Oxygen

b) Carbon dioxide

c) Nitrogen

d) Helium

149. Which blood cells fight infections?

a) Red blood cells

b) White blood cells

c) Platelets

d) Plasma

150. Which property of a substance allows it to be drawn into wires?

a) Malleability

b) Ductility

c) Conductivity

d) Elasticity

Answers

1. c) Root

2. c) Photosynthesis

3. c) Proteins

4. a) Digestion

5. c) Wool

6. c) Polyester

7. a) Temperature

8. b) Convection

9. b) 7

10. c) Hydrochloric acid

11. c) Melting of ice

12. b) Chemical change

13. b) Climate

14. a) Greenhouse gases

15. a) Polar bear

16. c) Anemometer

17. b) Lowpressure systems

18. c) Rocky soil

19. a) Minerals

20. b) Respiration

21. c) Alveoli

22. b) Xylem

23. b) Red blood cells

24. c) Reproduction

25. c) Stamen

26. c) Ovary

27. d) Meter per second

28. b) Motion

29. c) Second

30. a) Battery

31. b) It becomes a magnet

32. b) They repel

33. c) The Sun

34. d) Virtual and upright

35. c) Wind

36. b) Evaporation

37. c) 2.5%

38. a) To produce oxygen

39. a) Topsoil

40. b) Wastewater

41. a) Nitrogen

42. d) Vitamin D

43. a) Capillary action

44. c) Pedology

45. b) To exchange gases

46. b) Freezing water

47. d) Ampere

48. b) Refraction

49. b) Gravitational force

50. c) Burning paper

51. b) Potential energy

52. b) Controls cell activities

53. c) Natural gas

54. b) Transpiration

55. a) Cerebrum

56. c) Condensation

57. a) H_2O

58. c) Carbon dioxide

59. c) Loamy soil

60. b) Absorbs sunlight

61. a) Generator

62. c) Skull

63. c) Emission of sulfur dioxide and nitrogen oxides

64. d) Solar energy

65. c) Carbon dioxide

66. b) Carry oxygen

67. b) Weathering

68. c) Heart

69. b) Absorb water and nutrients

70. c) Copper

71. c) Oscillatory motion

72. c) Inertia

73. c) The Sun

74. b) Photosynthesis

75. c) Ovary

76. c) Oxygen

77. c) Na

78. b) Physical change

79. c) Bear

80. b) Thermometer

81. b) Newton

82. c) Clot blood

83. b) Refraction

84. b) Evaporation

85. b) Transpiration

86. c) Natural gas

87. b) Evaporation

88. c) Absorb water and form feces

89. c) Cell

90. c) Nitrogen

91. a) Biodiversity

92. b) Nucleus

93. b) Photosynthesis

94. c) Copper

95. b) Atom

96. b) Clay soil

97. c) Perform photosynthesis

98. c) Deforestation

99. b) Germination

100. a) Lever

101. c) Melting of ice

102. c) Carbon dioxide

103. b) Phloem

104. b) Asexual

105. c) Newton

106. b) Liver

107. b) Carbon dioxide

108. c) Large intestine

109. c) Copper

110. c) Distillation

111. c) Iris

112. b) Carbon dioxide

113. c) Vitamin C

114. c) Help in clotting

115. c) Radiation

116. c) Convex mirror

117. b) Chlorophyll

118. c) Ovary

119. b) Carbon dioxide

120. b) Diamond

121. b) Kinetic energy

122. c) Evaporation

123. b) Venus

124. b) Ampere

125. c) Citric acid

126. c) Semicircular canals

127. c) Jupiter

128. b) Nitrous oxide

129. a) Fermentation

130. b) Chemical energy

131. b) Kidney

132. b) Non-contact force

133. c) Coal

134. c) Leaves

135. a) Mercury

136. b) Stapes

137. b) Pasteurization

138. a) Liver

139. b) Saturn

140. b) Peregrine falcon

141. b) Barometer

142. b) Convex lens

143. a) Oxygen

144. a) Sublimation

145. c) Mitochondria

146. b) Neptune

147. b) Stem

148. b) Carbon dioxide

149. b) White blood cells
150. b) Ductility

9 789369 722709

Printed at Repro India Ltd.